Using Base Ten Materials To Teach

ADDITION AND SUBTRACTION

of 2-Digit Numbers

Connecting
Manipulatives
to　↓
Symbols
in Mathematics

Frank Gardella

innovativeink

PUBLISHING

A Division of Kendall Hunt

Cover image © Shutterstock.com

www.innovativeinkpublishing.com
Send all inquiries to:
4050 Westmark Drive
Dubuque, IA 52004-1840

Copyright © 2024 Francis Gardella

Print ISBN: 979-8-3851-2704-7
Ebook ISBN: 979-8-3851-2705-4

Published in the United States of America

Contents

Teaching Subtraction of 2-Digit Whole Numbers Using Base-10 Materials

Substraction: Trading Tens for Ones

Introduction

The use of physical models is extremely helpful in assisting students at all levels understand the mathematics that they are studying. The goal of instruction is to have the students use this understanding to work with mathematics at a symbolic level.

This book will assist you as you guide your students through the process of using what they learn from modeling mathematics with Base Ten materials to the recording of the mathematics using the standard mathematical symbols and algorithms for the Whole Number Operations of Addition and Subtraction.

What Are Base 10 Materials?

Base 10 Materials are models that can physically represent values in our place value system.

FOUR BASIC MODELS

There are four types of pieces in a set of **Base 10 Materials** when dealing with Whole Numbers:

i. A cube ⬜ which represents a value of '1'. This is called a 'one.'
ii. A bar of 10 cubes ▭▭▭▭▭ which represents a value of '10.'
    This is called a 'ten.'
iii. A flat of 100 cubes ▦ which represents the value of '100.' This is called a 'hundred.'
iv. A large cube or block containing 1000 cubes ▦ which represents the value of '1000.' This is called a 'thousand.'

Since the 'thousand cube' is in 3-dimensions, we cannot extend this type of geometric representation of any place value higher than 'thousands.'

NOTE: However, there are materials, such as Chip Trading developed by Patricia Davidson which use colored chips to represent place values. Using these materials, the place values can be extended beyond thousands.

REPRESENTING QUANTITIES

To represent the number 27, you can use 27 cubes as seen below.

or you can use 2-tens and 7-ones.

In using these materials, we can allow students to show numbers using any combination of the pieces. However, when we begin to link the pieces to the mathematical symbols, we encourage students to use as few pieces as possible. As you will see in the following pages with **Base 10 Materials** being used to model the operations of 2-digit numbers, it is best to have students use the **least number of pieces**. So, in the case of 27, the students would use 2-tens and 7-ones instead of 27-ones. This idea can be called **The Rule of Least Pieces** (Gardella, 2008) which stems for the work of Matthew Scaffa and Janet Castellano who were the Mathematics Supervisory Team for the New York City Public Schools on Staten Island in the 1970's through the 1990's.

The idea of "Least Pieces" gives a physical reason for the value in any place in our system never being more than 9.

Reference:

Gardella, F. (1996). *Least pieces: A Manipulative Necessity*. The New Jersey Mathematics Teacher, 54 (1), 16-19

Some History of Physical Models in Mathematics

Materials to physically model ideas of number have been used since people began using stones to represent quantities before symbolic numerals were developed. While the ideas of learner involvement stand prominently in education through the work of many leaders such as Jean Piaget, Lev Vygotsky and Jerome Bruner, the modern classroom use of physical models in learning mathematics can be traced to three people who worked in the same time-period in the middle of the 20th Century. These people were Zoltan Dienes, a mathematician, Catherine Stern, a Montessori-trained educator and George Cuisenaire, a musician and educator.

Zoltan Dienes (1916-2014) was born in Hungary but studied and conducted his early teaching in England. While teaching mathematics at the University of Leicester, he also attended courses in the education department and earned a Diploma in Education there in 1953. This led him to study Supplementary Psychology at the University of London. As he moved along in his career, he felt more and more that the mathematics that he saw as beautiful and exciting was, through being taught incorrectly, seen by most as scary and boring. It was the connection between his background in mathematics and psychology that had him begin to investigate how this connection could assist in the learning of mathematics.

Through the influence of Emile Borel, Dienes came to recognize how the personality of children impacted on their conceptualization of mathematics and that constructive thinking on the part of children was much more of a characteristic of children than the analytical thinking usually associated with learning mathematics. With this, he proposed the idea that teaching materials used should be based on the mathematical subject addressed rather than students' developmental stages based on their ages. This was in concert with his idea that by using manipulative materials, games and stories, children can understand more complicated mathematics at a younger age than had previously been thought.

Through his work as a member of the International Study Group of Mathematics Learning, he led research in the learning of mathematics by children between six and twelve years old.

Throughout his life, Dienes seemed to balance his work in mathematics with the way children learn mathematics and worked with elementary mathematics teachers wherever he was located.

This is why there are those who refer to Base-10 materials as Dienes Bars.

Catherine Stern (1894-1973) was a teacher and administrator in Breslau Germany in the late 1920's. She began to view the teaching of mathematics in a more structural fashion. While the Montessori Mathematics Materials used beads to make strings (such as 4 beads and 5 beads yielded a string of 9 beads), it made counting the individual beads the primary method for obtaining an answer of 9. Stern saw this differently. She glued 1-cm cubes together in a row and created linear number blocks in different colors. So, a block of 4 cubes and a block of 5 cubes was equal to a block of 9 cubes.

Emigrating to the United States in 1938, she introduced her materials at the Winward School in White Plains, NY. A major difference in her work here was that she used 1-inch cubes for the number bars as opposed to the 1-cm cubes used in Europe. As a result, in the United States today, these inch-bars are recognized as Stern Blocks.

Through her work with the psychologist Max Wertheimer, she realized that rote learning did not give learners insight into relationships. She saw that the visualization of the structural characteristics of the concept involved provided the true relationships of concepts that was needed by the learner.

Together with her daughter, Toni Gould and a student from Bank Street College, Margaret Bassett, the work of Catherine Stern was expanded to create published works that then became part of her Structural Arithmetic Program. A noted professional work was the book, Children Discover Arithmetic, co-authored with Margaret Bassett Stern. This writing sets the foundation for her work in helping children learn mathematics.

Her work addresses number ideas for simple numbers as well as the use of multiple materials in dealing with operations with multi-digit numbers.

More information relative to her program, called Stern Math, can be found at http://www.sternmath.com

Georges Cuisenaire (1891 – 1975) studied the violin and obtained both his license to teach school as well as music. He spent his professional life teaching in his native Belgium in Lower Town (Brussels) and then as Director of Primary Education. As a teacher he had a belief that, because of a child's affinity to color, the frequency ratios that exist in sounds can be developed using colors. In 1945 he began to produce sets of colored cardboard strips which he found very useful for learning arithmetic. Like Catherine Stern, he used definitive sized centimeter lengths to determine a number. In 1952, he produced his first and most well known work, "Les Nombres en Couleurs" (Numbers in Colour). He then met the psychologist, Caleb Gattegno who helped make the use of rods in teaching mathematics more widely known. Gattegno also saw that the rods allowed children to investigate aspects of mathematics on their own. Together, they published an account of the rods in English which led to the rods coming to the United States. In his work, Cuisenaire was also supported by Georges Papy who along with his wife, Frederique, was influential in the development of the Comprehensive School Mathematics Project by Fred Kaufman in the 1960's. This program is archived at the University of Buffalo.

Today, Cuisenaire rods are one of the most noteworthy materials in the teaching of mathematics throughout the world.

Certainly, Zoltan Dienes, Catherine Stern and George Cuisenaire are not the only mathematics educators who have interpreted number with models. However, they form in a sense a 'Founding Triumvirate' that certainly laid the foundation for what we use today in the teaching and learning of mathematics.

References:

For Zoltan Dienes

https://mathshistory.st-andrews.ac.uk/Biographies/Dienes_Zoltan

The history of base-ten-blocks: Why and who made base-ten-blocks? Rina Kim, Lillie R. Albert, Mediterranean Journal of Social Sciences, vol. 5, no. 9, pp. 356-365, May 2014

For mathematician and teacher Zoltan Dienes, the play was the thing Allison Lawlor, Globe and Mail, Feb. 4, 2014

For Catherine Stern

https://www.encyclopedia.com/women/encyclopedias-almanacs-transcripts-and-maps/stern-catherine-brieger-1894-1973

https://web.archive.org/web/20180406133248/http://www.sternmath.com/who-we-are.html

For George Cuisenaire

https://americanhistory.si.edu/teachingmath/html/302.htm

https://www.froebelweb.org/web2026.htmlboo

http://famousbelgians.net/cuisenaire.htm

file:///C:/Users/fgard/Downloads/ICHME6_DeBock_V3_reluGM+EV.pdf

https://www.eimacs.com/blog/2012/01/georges-papy-mathematics-educator-gifted-math-curriculum/

Levels of Learning: The Impact of Modeling

In using physical models in learning mathematics there are three levels in the process that students need to go through to truly understand mathematics and the symbolic system by with it is represented.

Concrete Level - What the students do using physical models

At the Concrete Level, the students use only the Base Ten materials to understand mathematics and to solve problems. There is no writing except for writing answers in complete sentences. In this initial step of addressing mathematics with Base Ten materials, the students build an understanding that at the next level will help them realize how to record their work and what the symbolic representations of mathematics really mean.

Transitional Level – How students learn to record the work they are doing with the Base Ten materials by using the symbols of mathematics

At the Transitional Level, the students learn how to take the mathematics that they have learned by modeling with Base Ten materials and communicate what they have done by recording their work using mathematical symbols. In this way, students learn what the symbols as well as their placement really are saying about mathematics.

Symbolic Level – Students use only mathematical symbols to do the work.

The Symbolic Level is the one that most adults remember learning in school. Many times, we see this as 'the mathematics' as we may have forgotten what was done in our learning prior to the development of the symbolism. It is at this level that students are free to express their work in writing, using the Base Ten materials only to verify their ideas if necessary.

It is important for students to move through the Concrete stage of working with Base Ten materials so they can begin to focus on solving problems before the direct use of the symbols.   In a sense, this work gives the students a view of the Base Ten materials to be stored in their 'mind's eye' for use when they will solve problems without the need for Base Ten materials. It is in this way that students learn the symbolic system by which mathematics is communicated.

About This Book

This book focuses on a system by which you as the teacher can assist the students as they transition from the use of Base Ten materials to communicating the work through writing with mathematical symbolism. Here, the focus is working with two operations with whole numbers, Addition and Subtraction.

Each of these operations is treated in their own section so that the teacher can focus on that single topic. For each operation, the text is separated into two parts.

<u>Part I</u> shows how the Base Ten Materials are used to solve an actual story problem. In this, the problem is solved using the Base Ten materials with diagrams showing what the students should be doing with the

materials. This is followed by a set of Teacher Practice Problems for you, the teacher, to help you make sure that you understand the process involved before bringing it to your students. And lastly, to help with the implementation of these ideas in your classroom, an outline of a lesson plan for the **Initial Teaching Lesson** is given. As part of the lesson plan, it is suggested that the Teacher Practice Problems be part of an activity during the class for students.

<u>**Part II**</u> of the writing shows the direct connection between the work with the Base Ten materials and the writing of the algorithm for the operation. In Part II, the same problem from Part I is again used as a model to enhance the connections involved. As each part of the problem is addressed using Base Ten materials, there is an accompanying diagram to show how the work is written using symbols.

Overview of the Book:

The focus of the book is to help you, the teacher, understand the place that physical models (manipulatives) have in mathematics teaching by:

a. Showing you through problem solving how Base Ten materials can be used by your students to understand mathematics.
b. Showing you how the models can lead to the symbolic representation of mathematics.
c. Giving you practice in each of these areas to be ready for your work in your classroom.
d. Giving you an outline of an **Initial Teaching Lesson** to guide you as you begin to use models with your students.

To review, below is the format for each section which focuses on a whole number operation.

Part I:

a. **A story problem will be given and solved using only Base Ten materials and a Base 10 Mat on which the students can work.**
b. **Several Teacher Practice Problems will be given to help you better understand the ideas of solving problems using Base Ten materials.**
c. **For use in your classroom teaching, an outline of an Initial Teaching Lesson will give you the information as to how to implement problem solving with Base Ten Materials in your classroom.**

Part II:

d. **The same story problem used in Part I will again be given and solved. Here the link between the use of Base 10 materials and the writing (recording) of what occurs will be shown.**
e. **The same Teacher Practice Problems from Part I will be given for you to better connect the work of solving problems using Base Ten materials to writing the symbolism and algorithm for that operation.**
f. **For use in your classroom teaching, an outline of an Initial Teaching Lesson will give you the information as to how to implement the connection between the use of Base Ten Materials to the written symbolism and algorithm for that operation. To assist with this, a Recording Sheet is included for use by the teacher in the modeling and by students in later work.**

Teaching Addition of 2-Digit Whole Numbers Using Base-10 Materials

Addition: Trading Ones for Tens

PART I CONCEPTUAL DEVELOPMENT

In addressing how to solve addition problems with models the first idea is to demonstrate how the students develop the concepts that underpin the addition of two 2-digit numbers with regrouping, or as we call it here, trading. The activity here allows students to physically work with the quantities and see the rationale for the need for trading and what happens to the ten that they trade for.

As with any problem solving, it begins with a story. This story will help you understand the concepts involved as well as form the basis for the Initial Teaching Lesson that is provided at the end of Part I.

The Story

Claire played two games at the park. The first time she played, she scored 27 points. The second time she played, she scored 35 points. When she told her friend, he said, "That's great. How many points did you score altogether?" This made Claire think of how to find this out.

Modeling the Story.

To model the story Claire used base ten materials.

For the first game she used 2-tens and 7-ones to show the 27 points.

Here is how Claire made a model on a Base-10 Mat using her materials.

as she showed 27 using 2-tens and 7-ones.

MAT A

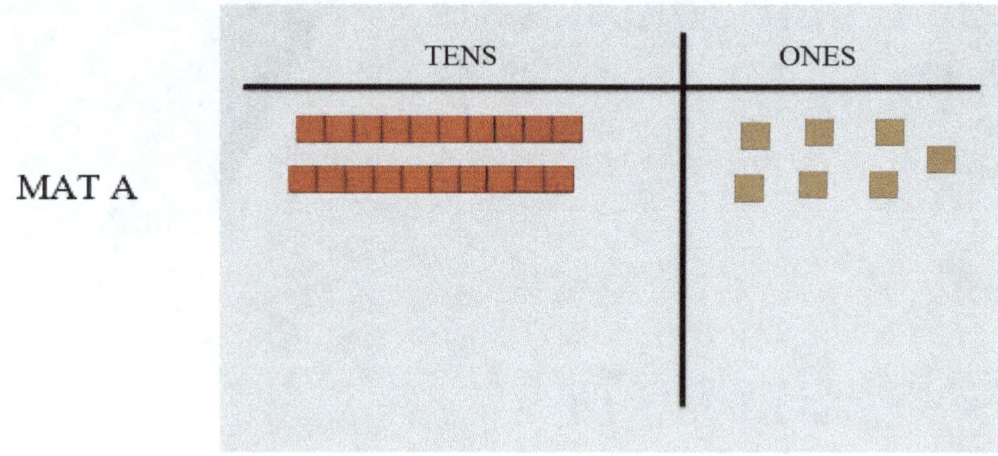

Then below 27 she showed her score in the second game, 35, as 3-tens and 5 ones.

MAT B

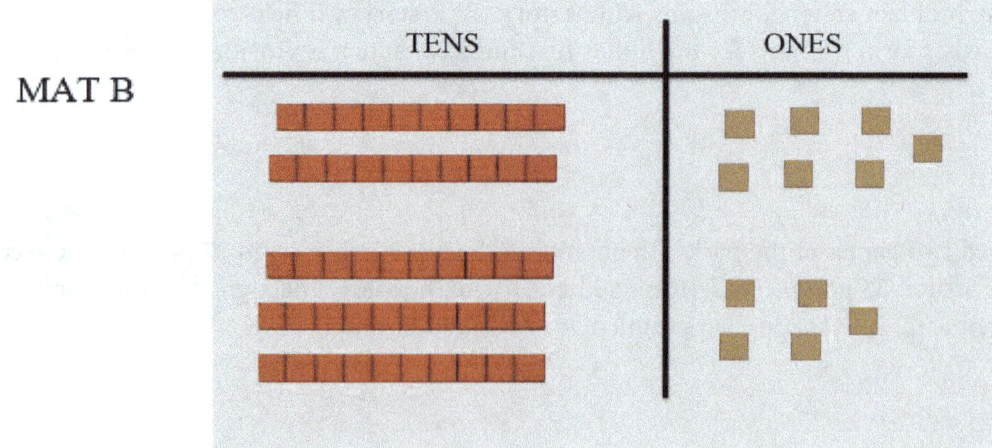

USING BASE TEN MATERIALS TO TEACH ADDITION AND SUBTRACTION OF 2-DIGIT NUMBERS:

Putting The Ones Together

Claire now put all the ones together.

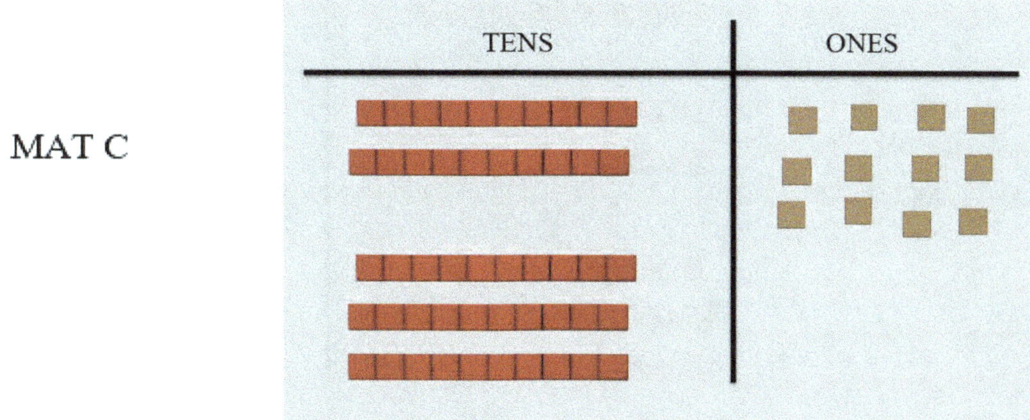

When she counted them, she found that she had 12-ones.

Making a New Ten

To have less pieces on the mat, she decided to trade 10-ones for a new ten.

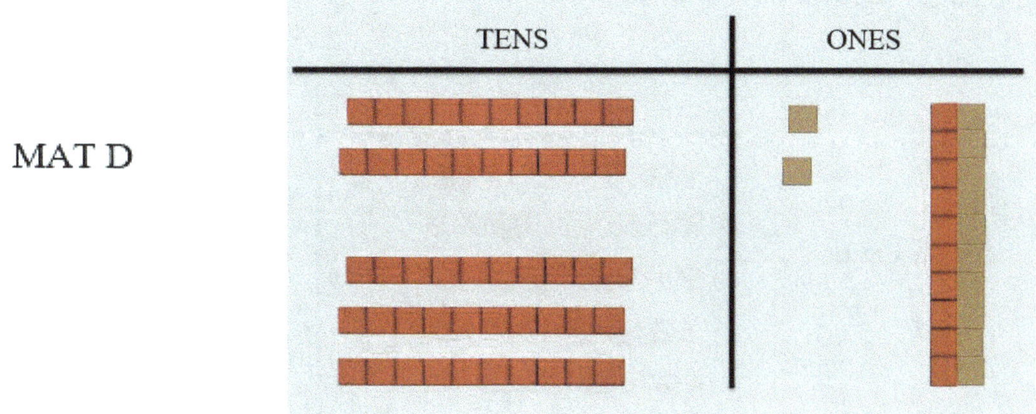

She then took away the 10-ones and put the new ten above the other tens.

This left her with 2-ones.

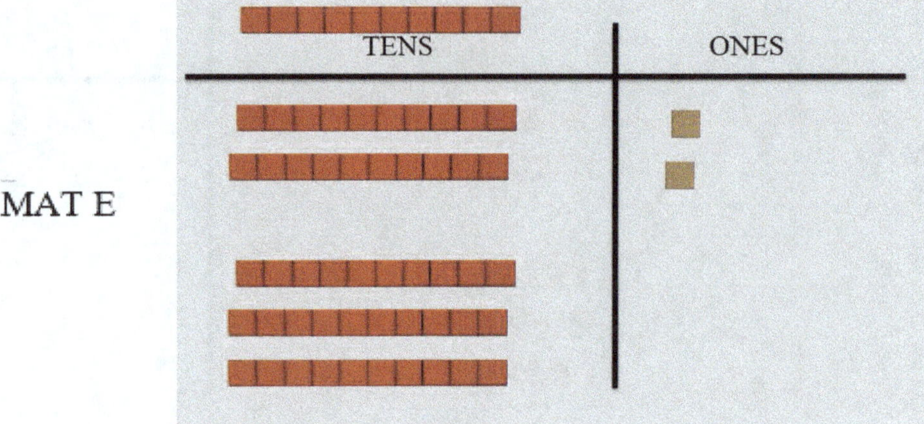

Putting the Tens Together

Claire now put the tens together.

From her initial work, she had 2-tens and 3-tens which gave her 5-tens.

She put these with the new ten.

When she put them together, she had 6-tens.

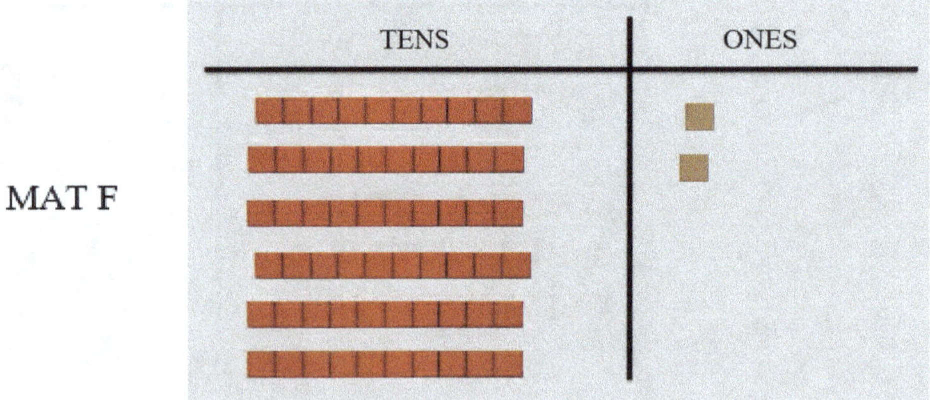

So, Claire now had 6-tens and 2-ones on her mat.

This equals 62.

Claire now knew that she had scored 62 points.

USING BASE TEN MATERIALS TO TEACH ADDITION AND SUBTRACTION OF 2-DIGIT NUMBERS:

Teacher's Practice: Here are some problems for you to try using only the base ten materials. The reason for these exercises is to have you assess your knowledge of their use and your ability to use them to solve problems. For these, concentrate on each of the MATS that you produce and the reasoning behind them. This will help in the next section when you review a sample Initial Teaching Lesson Plan that you may wish to use in your classroom. Lastly, to be a good model for your students, write the answer to each problem in a complete sentence.

1. In two days of basketball games, Thomas scored 54 points on day 1 and 38 points on day 2. How many points did Thomas score in the two days?

2. On a trip with her soccer team, Amy travelled 35 miles in the morning. After the game, they took a longer trip of 47 miles to return home. How many miles did Amy and her team travel?

3. In Samuel School, there are two second grade classes. Ms. Olivio has 19 students in her class. Mr. Hammond has 24 students in his class. How many students are there in second grade in Samuel School?

4. Walking from his home, Justin counted 26 cars parked on 15th Street. When he turned the corner to walk to school, he counted 18 cars parked on Wilmette Street before he arrived at school. How many parked cars did Justin count?

OUTLINE OF AN 'INITIAL TEACHING LESSON'

Objective: To have students work with Base Ten materials to solve a problem involving the addition of two 2-digit numbers using trading.

Materials: For each student, provide a set of 9 ten-bars and 19 ones.

A Place Value Mat. (A Model is given on the page after this lesson outline.)

Lesson Outline:

I Read or display the first two sentences of the story.

'Claire played two games at the park. The first time she played, she scored 27 points.'

Ask the students to show '27' on their Mats. (See **MAT A** above.)

After they do this, discuss why they should have put 2-tens and 7-ones on their Mat. (In doing this you do not have to relate the Rule of Least Pieces. Have the students know that using the Rule, it is easier to understand the numbers they are working with.)

II Now read or display the third sentence of the story.

'The second time she played, she scored 35 points.'

Have the students show this amount on their Mats. (See **MAT B** above.) Ask them to place this amount below the 2-tens and 7-ones that are already on the mat.

After they do this, discuss why they should have put 3-tens and 5-ones on their Mat.

III Now read or display the fourth sentence of the story.

'When she told her friend, he said, "That's great. How many points did you score altogether?"

Discuss with the students their ideas on how to show the total of all the points.

Then have them work in small groups (3 students) or individually to find the total number of points using the Base Ten materials.

Note: Remind the students that it is okay if they need to trade 10-ones for 1-ten.

IV After the students have worked with their ideas, have them decide on an answer. At their places, ask them to write the answer in a complete sentence.

V Have a student use either Base Ten materials and a document camera or a set of Base Ten materials for the chalkboard to explain what the student did.

VI Discuss the student's work and have others show their methods.

VII If you wish, when this is done, show the students how 'you would do it' using the explanations and **MAT C** through **MAT F** above. (In doing this, it will give the students a basis for the later work of linking the use of the Base Ten materials to the writing of the algorithm.)

VIII The students can be given the problems from the Teacher's Practice section above or from the text-book/program that is in use in the classroom. During this, the students will complete the work using the Base Ten materials and write each answer in a complete sentence. For some of the problems, the students can work in small groups while for other problems, you may wish to have them complete the work individually as a type of formative assessment.

Student Tens-Ones Mat

ONES

TENS

USING SYMBOLS TO RECORD THE WORK

SOME HISTORY ABOUT ADDITION SYMBOLS

The Symbol for Adding

While cultures have been adding numbers for a long time, the addition sign first began to be used in arithmetic in the 14th Century in the work of Nicole Oresme.

The sign itself seems to be a simplification of the Latin word, 'et' which means 'and.'

In more modern mathematics, the '+' sign was used by Henricus Grammateus in 1518.

Robert Recorde introduced the '+' sign to Britain in 1557. Interestingly, Recorde was also the designer of the 'equals' (=) sign.

So, with this information, let's begin to use Base 10 Materials to solve problems and then move on to record the work remembering that mathematical symbolism was and continues to be created by people.

PART II CONCEPTUAL AND SKILL DEVELOPMENT

Recording Our Work to Communicate Mathematics

Part II demonstrates the link between the work using the Base Ten materials and how this work can be recorded in ways that will allow the mathematics to be communicated to others.   The writing here shows how the physical use of the Base Ten materials is recorded on paper using the symbolism by which mathematics is communicated.

As with any problem solving, it begins with a story. The work solving the story will help you understand how the work with Base Ten materials is linked to the algorithm for addition of two 2-digit numbers. This will also form the basis for the Initial Teaching Lesson that is provided at the end of Part II.

The Story

Claire played two games at the park. The first time she played, she scored 27 points. The second time she played, she scored 35 points. When she told her friend, he said, "That's great. How many points did you score altogether?" This made Claire think of how to find this out. She decided to make a model and find the answer.

Modeling the Story.

To model the story, Claire used base ten materials. For the first game she used 2-tens and 7-ones to show the 27 points. For the second game, she used 3-tens and 5-ones to show the 35 points.

Here is how Claire made a model using her materials on a Base-10 Mat.

First, she showed 27 using 2-tens and 7-ones.

She then recorded her work on a Tens/Ones chart writing a '2' in the Tens column and a '7' in the Ones column.

MAT A

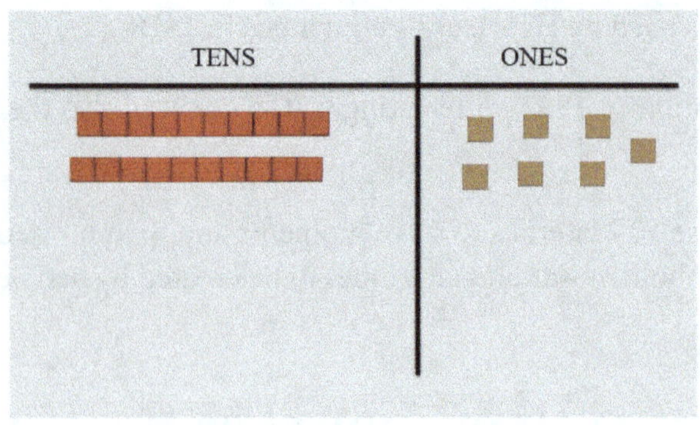

LINK A

TENS	ONES
2	7

Then below 27 she showed 35 as 3-tens and 5 ones.

She recorded 35 by writing a '3' in the Tens column and a '5' in the Ones column.

MAT B

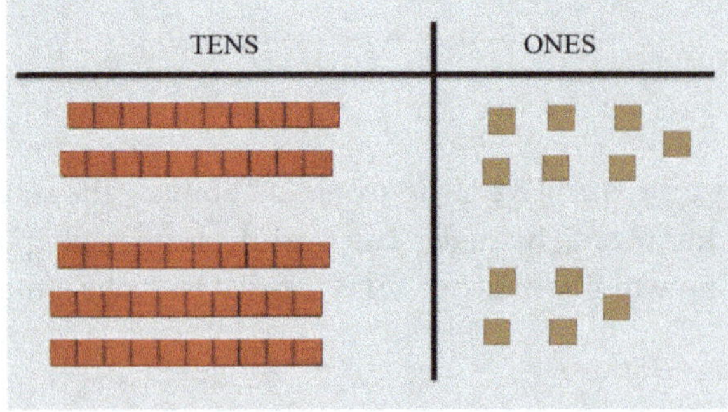

LINK B

TENS	ONES
2	7
3	5

Since Claire knew that she was going to add these numbers, she put a "+" sign next to the 3. She also put a line under the '3' and '5' to show that anything under the line would be the total points that she scored.

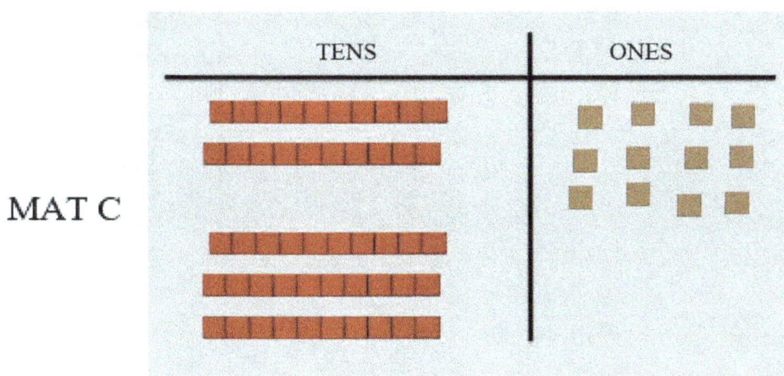

MAT B

LINK B

TENS	ONES
2	7
+ 3	5

Putting The Ones Together

Claire now put all the ones together.

MAT C

LINK C

TENS	ONES
2	7
+ 3	5

When she counted them, she found that she had 12-ones.

Making a New Ten

To have less pieces on the mat, she decided to trade 10-ones for a new ten.

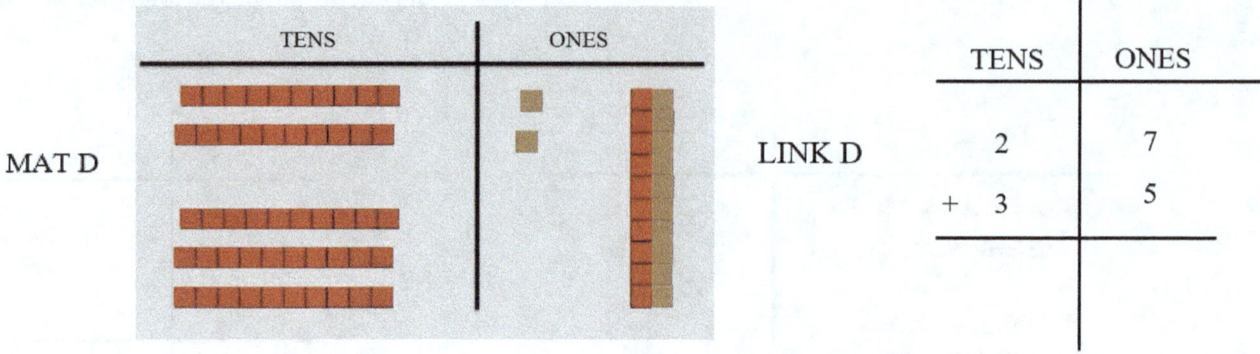

MAT D

LINK D

	TENS	ONES
	2	7
+	3	5

She then took away the 10-ones and put the new ten above the other tens.

She recorded the new Ten by writing a '1' above the '2' in the Tens column.
This left her with 2-ones. She wrote '2' in the Ones column under the line.

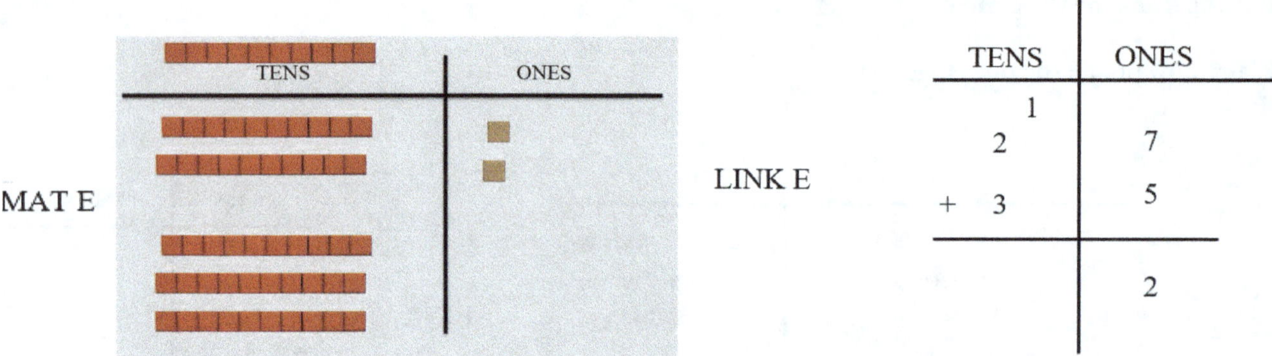

MAT E

LINK E

	TENS	ONES
	1	
	2	7
+	3	5
		2

Putting the Tens Together

Claire now put the tens together.

She had 2-tens and 3-tens which gave her 5-tens.

She added these to the new ten.
This gave her 6-tens.

She recorded a '6' in the Tens column
under the line.

MAT F

TENS	ONES
1	
2	7
+ 3	5
6	2

LINK F

So Claire now had 6-tens and 2-ones. This equals 62.

Claire scored 62 points in the two games.

Teacher's Practice: Here are some problems for you to try using base ten materials and relating them to the writing of the algorithm. The reason for these exercises is to have you assess your knowledge of moving from Base 10 materials to the symbolism in solving problems. In completing these problems, focus on each of the MATS that you produce, the reasoning behind them and their linkage to the writing of the algorithm for addition of 2-digit numbers. This will help in the next section when you review a sample Initial Teaching Lesson Plan that you may wish to use in your classroom to help your students establish this linkage. Again, to be a good model for your students, write the answer to each problem in a complete sentence.

(Please note: These problems are the same as those from Part I to allow you to focus on the important linkage of the Base Ten materials and the written algorithm.)

1. In two days of basketball games, Thomas scored 54 points on day 1 and 38 points on day 2. How many points did Thomas score in the two days?
2. On a trip with her soccer team, Amy travelled 35 miles in the morning. After the game, they took a longer trip of 47 miles to return home. How many miles did Amy and her team travel?
3. In Samuel School, there are two second grade classes. Ms. Olivio has 19 students in her class. Mr. Hammond has 22 students in his class. How many students are there in second grade in Samuel School?
4. Walking from his home, Justin counted 26 cars parked on 15th Street. When he turned the corner to walk to school, he counted 18 cars parked on Wilmette Street before he arrived at school. How many parked cars did Justin count?

OUTLINE OF AN 'INITIAL TEACHING LESSON'

Objective: To have students understand the link between the work they did with the Base Ten materials with Addition of two 2-digit numbers and how to record this work on paper.

Materials:

For each student:

a.   Provide a set of 9 ten-bars and 19 ones.
b.   A Place Value Mat. (Students can use the Mat from their previous work.)
c.   A Tens-Ones Recording Page for students to use in recording the symbols as they are developed in solving the problems.

For the teacher:

A Tens-Ones Recording Page for your use in recording the symbols as they are developed.   (A Model is given on the page after the lesson outline.)

<u>NOTE</u>: In this Initial Teaching Lesson, the teacher acts as the recorder while the students solve the problem using Base Ten materials. With subsequent problems, the students will also do the recording.

Lesson Outline:

I   Read or display the first two sentences of the story.

'Claire played two games at the park. The first time she played, she scored 27 points.'

Ask the students to show '27' on their Mats. (See **MAT A** above.)

After they do this, discuss why they should have put 2-tens and 7-ones on their Mat.

Then record **27** on the Tens-Ones Recording Page so the students see how their work is written.

(See **Link A** above.)

II   Now read or display the third sentence of the story.

'The second time she played, she scored 35 points.'

Have the students show this amount on their Mats. (See **MAT B** above.) Ask them to place this amount below the 2-tens and 7-ones that are already on the mat.

After they do this, discuss why they should have put 3-tens and 5-ones on their Mat.

Then record **35** on the Tens-Ones Recording Page below 27 to link to what the students have on their Place Value Mats.

(See **Link B** above.)

III Now read or display the fourth sentence of the story.

'When she told her friend, he said, "That's great. How many points did you score altogether?" '

Discuss with the students their ideas on how to show all the points.

Then have them work in small groups (3 students) or individually to find the total number of points using the Base Ten materials.

Note: Remind the students that it is okay if they need to trade 10-ones for 1-ten.

IV   After the students have worked with their ideas, explain why you put a '+' sign and a line under the numbers on the Tens-Ones Recording Page to indicate where the answer will be written. **Link C**

V Choose students individually to report on their work	Links to Writing
i Put the ones together to have 12 ones	**No recording for this**
ii Trade 10-ones for 1-ten	**No recording for this**
iii Leave 2-ones in the ones-place.	
Put New Ten at the top of the Tens column	**Link E**
iv Add the tens from the problem and then the new ten	**Link F**

VI   Begin again and have other students state what they did as you show the writing of the symbols.

VII The students can be given the problems from the Teacher's Practice section above or from the textbook/ program that is in use in the classroom. Students can work individually. Also, students can work in pairs where one student works with the Base Ten materials to solve the problem while the other records the work. After one problem, they can switch so that each student gets practice recording the work being done. For each problem, the students should decide on an answer and write it in a complete sentence.

TENS | ONES

Teaching Subtraction of 2-Digit Whole Numbers Using Base-10 Materials

Trading Ones for Tens

PART I CONCEPTUAL DEVELOPMENT

In addressing how to solve subtraction problems with models the first idea is to demonstrate how the students develop the concepts that underpin the subtraction of 2-digit numbers with regrouping, or as we call it here, trading. The activity here allows the students to physically work with the quantities and see the rationale for the need for trading and what happens to the ten that they trade for 10-ones.

As with any problem solving, it begins with a story. This story will help you understand the concepts involved as well as form the basis for the Initial Teaching Lesson that is provided at the end of Part I.

The Story

To create a design for art class, Laura bought a box which contained 62 small tiles. The box contained 6 packs of 10 tiles each and 2 other tiles for a total of 62 tiles. With this many tiles, Laura thought that she could make two designs. In making her first design, Laura planned to use 27 tiles. When she finished her first design, Laura wanted to know how many tiles she would have left to create a second design.

Modeling the Story.

To model this, Laura used base ten materials. Since there were 6 packs of 10 tiles and 2 other tiles in the box, she showed 6-tens and 2-ones on her Base 10 mat.

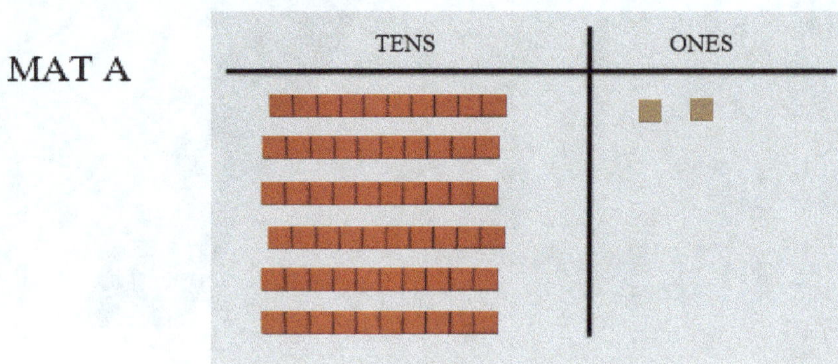

Taking Away the Ones

She wanted to take 27 tiles away from the 62 tiles. So, she began by trying to take away 7-ones.

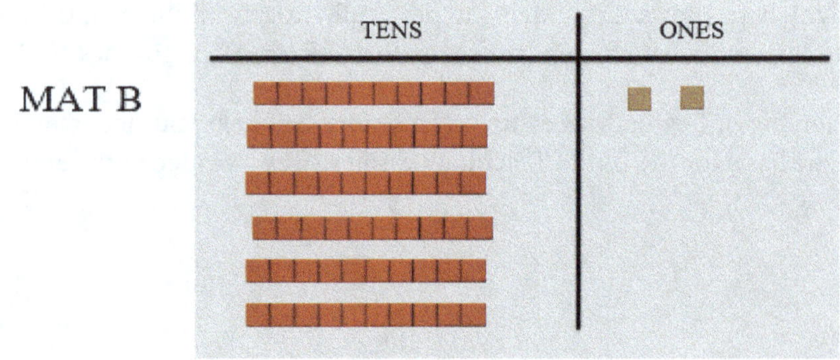

USING BASE TEN MATERIALS TO TEACH ADDITION AND SUBTRACTION OF 2-DIGIT NUMBERS:

But she could not take away 7-ones since she only had 2-ones.

So, she decided to trade 1-ten for 10-ones

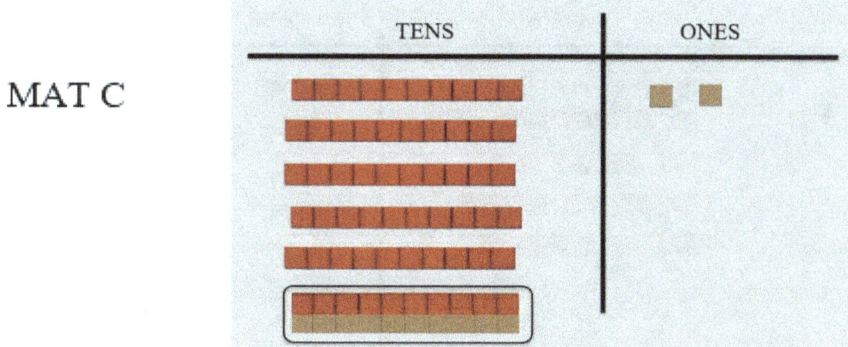

She then took the 1-ten off the mat and put the 10-ones in the ones column.

This left 5-tens in the tens place.

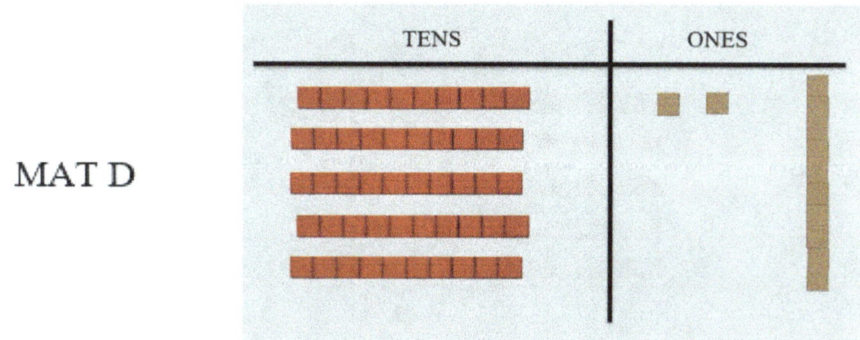

When she counted the ones, she found that she had 12-ones.

She now had enough ones to take away 7-ones.

When she did this, she had 5-ones left.

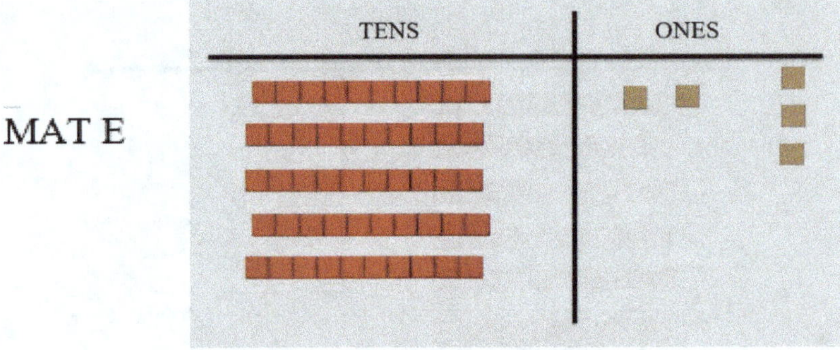

Taking Away the Tens

She then worked with the tens. She took 2-tens from the mat.

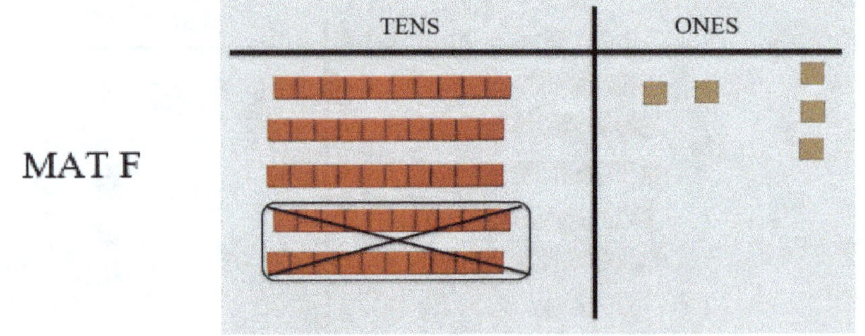

Laura now had 3-tens and 5-ones.

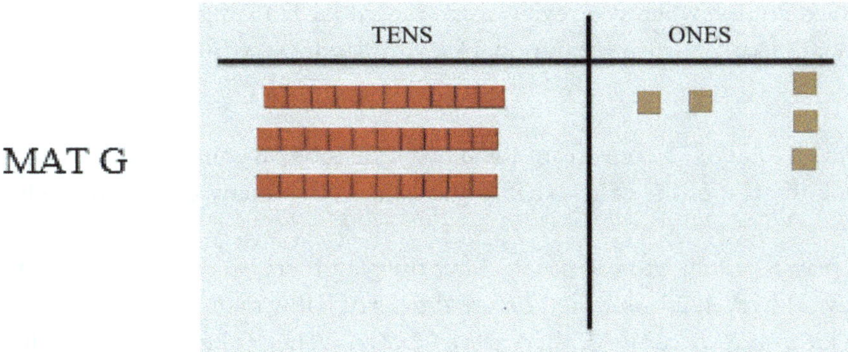

3-tens and 5-ones is equal to 35.

Laura now knew that she could create another design with 35 tiles.

Teacher's Practice: Here are some problems for you to try using only the base ten materials. The reason for these exercises is to have you assess your knowledge of their use and your ability to use them to solve problems. For these, concentrate on each of the MATS that you produce and the reasoning behind them. This will help in the next section when you review a sample Initial Teaching Lesson Plan that you may wish to use in your classroom. Lastly, to be a good model for your students, write the answer to each problem in a complete sentence.

1. Alicia collected a total of 73 coins from the bank. She took 36 coins and put them into her collection. She took the rest of the coins back to the bank. How many coins did she bring back to the bank?
2. For the class play, 52 adults came to see it. After the play, there was a party for the people who came to see the play. However, 26 adults left before the party. How many adults went to the party?
3. In a parking lot near a soccer field, there were 92 cars. When it began to rain during the game 37 cars drove out of the parking lot. How many cars are still in the parking lot?
4. There are 62 music students in the all-purpose room. Some of them are in the band and some are in the choir. Mr. Richmond takes 23 students and leads them to the band room. How many students are still in the all-purpose room?
5. Maria had 63 building blocks. She gave her little brother 28 to use. How many blocks did Maria have left?

OUTLINE OF AN 'INITIAL TEACHING LESSON'

Objective: To have students work with Base Ten materials to solve a problem involving the subtraction of two 2-digit numbers using trading.

Materials: For each student, provide a set of 9 ten-bars and 19 ones.

A Place Value Mat. (A Model is given on the page after the lesson outline.)

Lesson Outline:

I   Read or display the first two sentences of the story.

'To create a design for art class, Laura bought a box which contained 62 tiles.   There were 6 packs of 10 tiles and 2 other tiles for a total of 62 tiles.'

Ask the students to show '62' on their Mats. (See **MAT A** above.)

After they do this, discuss why they should have put 6-tens and 2-ones on their Mat.

II   Now read or display the next two sentences of the story.

'With this many tiles, Laura thought that she could make two designs. In making her first design, Laura planned to use 27 tiles.'

Discuss with the students what this means. (Need to take 27 from the 62 on the mat.)

(Note that there is no work with the materials. See **Mat B.** In this story, the 27 that she wants to remove are already on **Mat B.** The 27 tiles are contained in the 62 tiles. So there are no other materials to place on the mat.)

III   Now read or display the fifth sentence of the story.

"When she finished her first design, Laura wanted to know how many tiles she would have left to create a second design."

Discuss with the students their ideas on how to remove 2 tens and 7 ones.

Then have them work in small groups (3 students) or individually to find work with the idea of removing 2 tens and 7 ones.

Note: In the work here, we will be focusing on the removal of the 7 ones first. For this reason, remind the students that it is okay if they need to trade a ten for 10-ones.

IV   After the students have worked with their ideas, have them decide on an answer. At their places, ask them to write the answer in a complete sentence.

V   Have a student use either Base Ten materials and a document camera or a set of Base Ten materials for the chalkboard to explain what the student did.

VI   Discuss the student's work and have others show their methods.

VII   If you wish, when this is done, show the students how 'you would do it' by trading to be able to remove the 7 ones, using the explanations and **MAT C** through **MAT G** above. (In doing this, it will give the students a basis for the later work of linking the use of the Base Ten materials to the writing of the algorithm.)

VIII   The students can be given the problems from the Teacher's Practice section above or from the textbook/program that is in use in the classroom. During this, the students will complete the work using the Base Ten materials and write each answer in a complete sentence. For some of the problems, the students can work in small groups while for others you may wish to have them complete the work individually as a type of formative assessment.

ONES

TENS

USING SYMBOLS TO RECORD THE WORK

SOME HISTORY ABOUT SUBTRACTION SYMBOLS

The Symbol for Subtracting

The name 'minus' comes from the Latin language in which 'minus' means less.

In 15th century Europe, the letter, **M**, was used. In 1494, another 'minus sign,' **m**, was used by the mathematician Luca Pacioli in a work published in Venice, Italy.

Like the 'plus' sign, the 'minus' sign appeared as it is now used in a book by Henricus Grammateus in 1518.

Also, like the '+' sign, Robert Recorde introduced the '-' sign to Britain in 1557. As you may remember from before, Recorde was also the designer of the 'equals' (=) sign.

PART II CONCEPTUAL AND SKILL DEVELOPMENT

Recording Our Work to Communicate Mathematics

Part II demonstrates the link between using the Base Ten materials to solve a problem and how this work can be recorded in ways that will allow the mathematics to be communicated to others.   The writing here shows how the physical use of the Base Ten materials is recorded on paper using the symbolism by which mathematics is communicated.

As with any problem solving, it begins with a story. Solving the story will help you understand how the work with Base Ten materials is linked to the algorithm for subtraction of two 2-digit numbers. This will also form the basis for the Initial Teaching Lesson Plan that is provided at the end of Part II.

The Story

To create a design for art class, Laura bought a box which contained 62 tiles. There were 6 packs of 10 tiles each and 2 other tiles for a total of 62 tiles. With this many tiles, Laura thought that she could make two designs. In making her first design, Laura planned to use 27 tiles. When she finished her first design, Laura wanted to know how many tiles she would have left to create a second design.

Modeling the Story.

To model this, Laura used Base Ten materials.

Since there were 6 packs of 10 tiles and 2 other tiles in the box, she showed 6-tens and 2-ones on her tens/ones mat.

She recorded her work on tens/ones chart writing a '6' in the Tens-column and a '2' in the Ones-column.

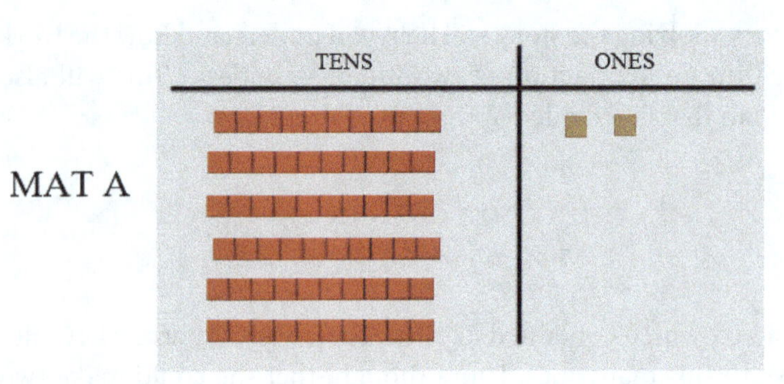

MAT A

LINK A

TENS	ONES
6	2

Taking Away the Ones

She wanted to take 27 tiles away from the 62 tiles.

This is the way she recorded her work.

MAT B

LINK B

TENS	ONES
6	2
- 2	7

So she began by taking away the ones.

But she could not take away 7-ones since she only had 2-ones.

MAT B

TENS	ONES
6	2
- 2	7

LINK B

So she decided to trade 1-ten for 10-ones

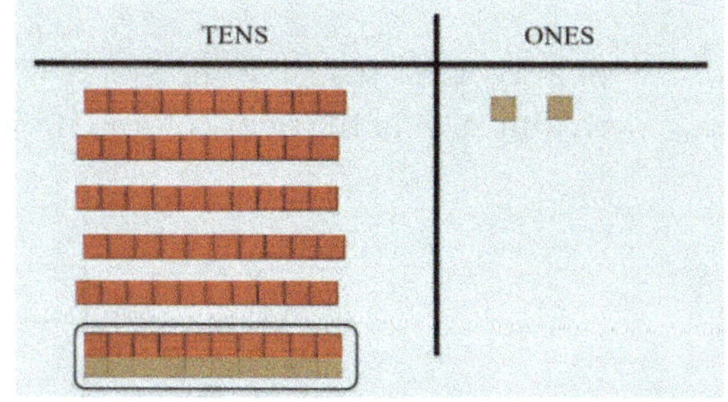

MAT C

TENS	ONES
6	2
- 2	7

LINK C

She then took the 1-ten off the mat and put the
10-ones in the one place.
When she counted the ones, she found that
she had 12-ones.
This left 5-tens in the tens-column.

She recorded this on the tens/ones chart.
In the ones column, she crossed out the '2'
and wrote '12' since she now had 12-ones.
In the tens column, she crossed out the '6'
and wrote '5' since she now had 5-tens.

MAT D

LINK D

TENS	ONES
5	12
6̸	2̸
- 2	7

She now had enough ones to take away 7-ones.
When she did this, she had 5-ones left.

She recorded that she had 5-ones left by
writing a '5' in the ones column under the line.

MAT E

LINK E

TENS	ONES
5	12
6̸	2̸
- 2	7
	5

Taking Away the Tens

She then worked with the tens.
She took 2-tens from the mat.
This left her with 3-tens.

To record this, she wrote '3'
in the tens-column under the line.

MAT F 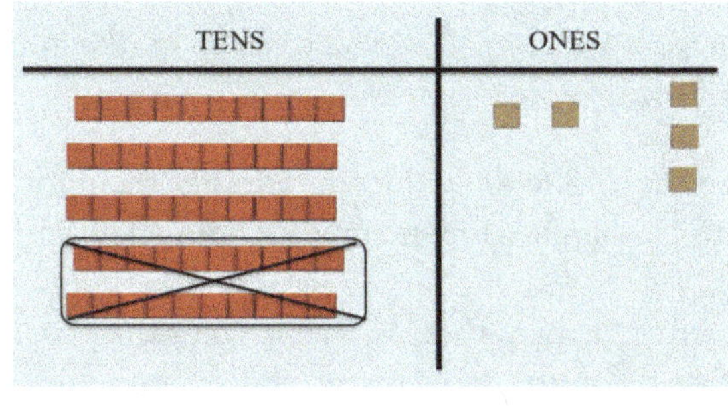

LINK F

TENS	ONES
5	12
Ø	2̷
- 2	7
3	5

Laura now had 3-tens and 5-ones on her tens/one mat.

On her tens/ones chart at the bottom, she had 3 in the tens column and 5 in the ones column.

MAT G 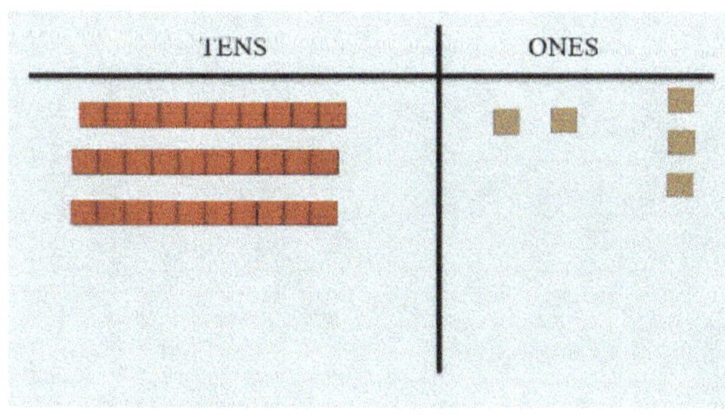

LINK G

TENS	ONES
5	12
Ø	2̷
- 2	7
3	5

3-tens and 5-ones is equal to 35.

Laura knew that she could make another design with 35 tiles.

Teacher's Practice: Here are some problems for you to try using base ten materials and relating them to the writing of the algorithm. The reason for these exercises is to have you assess your knowledge of moving from Base 10 materials to the symbolism in solving problems. In completing these problems, focus on each of the MATS that you produce, the reasoning behind them and their linkage to the writing of the algorithm for subtraction of 2-digit numbers. This will help in the next section when you review a sample Initial Teaching Lesson Plan that you may wish to use in your classroom to help your students establish this linkage. Again, to be a good model for your students, write the answer to each problem in a complete sentence.

(Please note: These problems are the same as those from Part I to allow you to focus on the important linkage of the Base Ten materials and the written algorithm.)

1.   Alicia collected a total of 73 coins from the bank. She took 36 coins and put them into her collection. She took the rest of the coins back to the bank. How many coins did she bring back to the bank?
2.   For the class play, 52 adults came to see it. After the play, there was a party for the people who came to see the play. However, 26 adults left before the party. How many adults went to the party?
3.   In a parking lot near a soccer field, there were 92 cars. When it began to rain during the game 37 cars drove out of the parking lot. How many cars are still in the parking lot?
4.   There are 62 music students in the all-purpose room. Some of them are in the band and some are in the choir. Mr. Richmond takes 23 students and leads them to the band room. How many students are still in the all-purpose room?
5.   Maria had 63 building blocks. She gave her little brother 28 to use. How many blocks did Maria have left?

OUTLINE OF AN 'INITIAL TEACHING LESSON'

Objective: To have students understand the link between the work they did with the Base

Ten materials with Subtraction of two 2-digit numbers and how to record this work on paper.

Materials:

For each student:

    a.   **Provide a set of 9 ten-bars and 19 ones.**
    b.   **A Place Value Mat. (Students can use the Mat from their previous work.)**
    c.   **A Tens-Ones Recording Page for students to use in recording the symbols as they are developed in solving the problems.**

For the teacher:

A Tens-Ones Recording Page for your use in recording the symbols as they are developed. (A Model is given on the page after the lesson outline.)

<u>**NOTE**</u>**: In this lesson, the teacher acts as the recorder while the students solve the problem using Base Ten materials.**

Lesson Outline:

I   Read or display the first two sentences of the story.

'To create a design for art class, Laura bought a box which contained 62 tiles.   There were 6 packs of 10 tiles and 2 other tiles for a total of 62 tiles.'

Ask the students to show '62' on their Mats.

After they do this, discuss why they should have put 6-tens and 2-ones on their Mat. On the Tens-Ones Recording Page, you write 62 in the proper places. **(Link A)**

II Now read or display the next two sentences of the story.

'With this many tiles, Laura thought that she could make two designs. In making her first design, Laura planned to use 27 tiles.'

Discuss with the students what this means. (Need to take 27 tiles from the 62 tiles.)

(Note that there is no work with the materials.) Write '-27' on the **Tens-Ones Recording Page (Link B)** along with a line underneath the 27.

III Now read or display the fifth sentence of the story.

"When she finished her first design, she wanted to know how many tiles she would have left to create a second design."

Discuss with the students their ideas on how to remove 2 tens and 7 ones that they worked with in Part I of this activity. As the students relate what they did, record this on the **Tens-Ones Recording Page (Links C to G)**

IV Review the story with the students, the work they did with the Base Ten materials to solve the problem and the recording that you did.

IV The students can be given the problems from the Teacher's Practice section above or from the textbook/program that is in use in the classroom. Students can work individually. Also, students can work in pairs where one student works with the Base Ten materials to solve the problem while the other records the work. After one problem, they can switch so that each student gets practice recording the work being done. For each problem, the students should decide on an answer and write it in a complete sentence.

For some of the problems, you may wish to have them complete the work individually as a type of formative assessment.

TENS	ONES